Carla Graciele Gomes

Sensitising students about Dasypodidae in Abelardo Luz - SC

Carla Graciele Gomes

Sensitising students about Dasypodidae in Abelardo Luz - SC

Hunting is one of the main causes of animal decline, which if not reduced could lead to extinction

ScienciaScripts

Imprint
Any brand names and product names mentioned in this book are subject to trademark, brand or patent protection and are trademarks or registered trademarks of their respective holders. The use of brand names, product names, common names, trade names, product descriptions etc. even without a particular marking in this work is in no way to be construed to mean that such names may be regarded as unrestricted in respect of trademark and brand protection legislation and could thus be used by anyone.

Cover image: www.ingimage.com

This book is a translation from the original published under ISBN 978-613-9-72159-7.

Publisher:
Sciencia Scripts
is a trademark of
Dodo Books Indian Ocean Ltd. and OmniScriptum S.R.L publishing group

120 High Road, East Finchley, London, N2 9ED, United Kingdom
Str. Armeneasca 28/1, office 1, Chisinau MD-2012, Republic of Moldova, Europe
Printed at: see last page
ISBN: 978-620-7-94128-5

I dedicate this work to my parents: Ademar and Fátima, for their unconditional support.

ACKNOWLEDGEMENTS

To my parents, Ademar José Gomes and Maria de Fátima Gomes, for all the effort they put in so that I could get this far. I will never be able to express all my gratitude. Thank you so much for your support, affection and trust, for believing in me and for never giving up. Especially my mum, who accompanied me on all the visits and talks to schools in the rural area of the city, rain or shine she was always by my side. I would also like to thank the rest of my family who have been close to me throughout this journey.

To my boyfriend Willian Matheus Derossi, for his help in correcting the questionnaires, for understanding when I was absent to do this work. And for always being by my side despite everything. Thank you so much for your strength and encouragement, I love you!

To my supervisor Professor Alan Miranda Prestes, for his professionalism, seriousness and intelligence. For his help and patience during all stages of the research, especially when it came to analysing the data and spreadsheets.

To my colleagues Gabriela Lorenzon, Daiana De Moura and Daniela A. B. de Oliveira for the conversations, laughs, outbursts and support, without you I would never have made it. Thank you, friends! To Adolfo for his friendship and support in my work.

To my co-supervisors Fernanda Góós Braga and Renata Bicudo Molinari, who even from afar gave me a lot of help with the writing and assembly of the questionnaire, especially Fer for the idea to carry out this project.

To Professors Elisangela Bini Dorigon and Francielle Garghetti Battiston, who provided all their help and support throughout the research.

Thank you to everyone who has been with me on this journey!

"Success is born of desire, determination and persistence in reaching a goal. Even if you don't reach the target, those who seek and overcome obstacles will at least do admirable things." (José de Alencar).

SUMMARY

Hunting, to this day, is one of the main causes of animal decline, whether for food or for sport, and consequently leads to extinction. The lack of studies in the area of animal conservation is also a problem to be solved in schools, so the aim of this study was to identify the perceptions of public school students in the municipality of Abelardo Luz/SC regarding the hunting and conservation of armadillos. The study was carried out with 128 ninth-graders from six public schools, through the application of a questionnaire before and after the presentation of a short lecture on the subject. Based on the students' previous knowledge, it was possible to observe that there is little knowledge about the species and that it suffers greatly from hunting, which is often carried out by the students' family and friends. It was noted that 10 per cent of the students interviewed do not know and/or have never seen the armadillo in their region or city, however, of all the animals available in the questionnaire, the armadillo predominated among the most hunted in the students' opinion. It was also possible to see that, after the explanatory talk, there was a 23 per cent reduction, from 26 per cent to just 3 per cent, in the students' opinion on the use of armadillos for food. As a result, the lecture method helped the students to understand the importance of species conservation, raising awareness among those interviewed and thus reducing the impact of hunting in the region studied.

Keywords: Biodiversity. Biology. Hunting. Lectures.

SUMMARY

CHAPTER 1

INTRODUCTION

Armadillos are the common name given to placental mammals that belong to the order Xenarthra and family Dasypodidae. They are characterised by having dermal bone plates that form the bony carapace (osteoderms) (NOWAK, 1999; WETZEL, 1985), as well as having simple or absent teeth in many cases, a fact that gave them the name Edentata (HILDEBRAND & GOSLOW, 2006). They are fossorial animals that use their claws to dig (BLOCH et al., 1976).

According to Wetzel (1985), armadillos comprise eight genera in the Dasypodidae family: Chlamyphorus (pichiegos), Zaedyus (pichi), Euphractus (tatu-peba), Tolypeutes (tatu-bola), Dasypus (tatu-galinha), Cabassous (tatu-de-rabo-mole), Chaetophractus (tatu- peludo) and Priodontes (tatu-canastra) and 21 recorded species. They are geographically distributed from the United States to Argentina, and are the largest family in the order (EISENBERG & REDFORD, 1999). In Brazil they are widely distributed, with five genera and 10 species, and one native species, the *Tolypeutes* tricinctus (ball armadillo) (PRADA & MARINHO-FILHO, 2004).

Anacleto (2006) mentions these armadillo species, which are studied in the six Brazilian biomes (Pantanal, Atlantic Forest, Caatinga, Pampa, Cerrado and Amazon): *Cabassous unicinctus* (common soft-tailed armadillo), *Cabassous tatouay* (*large* soft-tailed armadillo*),Euphractus sexcmctus* (peba armadillo), *Priodontes maximus* (canastra armadillo), *Tolypeutes matacus* (ball armadillo), *Tolypeutes tricinctus* (tatu-bola-da-caatinga), *Dasypus hybridus* (tatu- mulita), *Dasypus kappleri* (tatu-de-quinze-quilos), *Dasypus novemcinctus* (tatu-galinha or tatu-verdade), *and Dasypus septemcinctus* (tatu-galinha-pequinha or tatu-mirim).

Despite their ecological importance, armadillos are still targeted as food resources in rural areas in Brazil, even though this practice is prohibited by the country's environmental protection agencies (DEPS *et al.,* 2003). According to Law No. 5,197 of 3 January 1967, the persecution of any wild animal is prohibited, as is the use of these animals in captivity, hunting or for food.

According to Alves *et al.* (2012), hunting and other uses of wild fauna are ancient practices, dating back to the time when human-environment interaction was based on survival needs, such as obtaining food, clothing and protection. This practice has been maintained as a cultural activity in some communities, negatively impacting some of the species targeted and generating an ecological problem (ALVES *et al.* 2012).

In addition, there has also been an accelerated and large-scale expansion in the number of motorways built in the country, which is one of the main causes of biodiversity loss, not only because of the increase in road traffic, but also because they promote habitat fragmentation, edge effects, loss of genetic variability due to isolation of groups, among other factors (BAGER, 2012). One of the groups impacted by these developments is the Dasypodidae, which is even more threatened when you consider that they are also targets for hunting (DIAS, 2013).

The lack of studies in the area of animal conservation is also a problem to be solved in schools. As a result, there is no guidance for the population on the impacts that hunting has on our biodiversity. Because they reproduce what they hear in their environment and day-to-day life, children are often the expression of a community's culture and moral development. This is why it is important to analyse how they perceive the issue addressed here, in order to obtain indications of how their parents and the adults they live with deal with the subject. With this in mind, the aim of this work was to identify the perceptions of students from public schools in Abelardo Luz and the surrounding area regarding the hunting and conservation of armadillos.

CHAPTER 2

RESEARCH QUESTIONS

Do the students know the genus Dasypodidae? Does it exist in your region/city?

Do the students know anyone who hunts animals? If so, what animals do they hunt?

Knowing little about the subject, how do the students see this animal?

Why should this genre be preserved?Have students' perceptions changed?

CHAPTER 3

LITERATURE REVIEW

3.1ARMADILLO HUNTING IN BRAZIL

According to Costa Neto et al. (2000), the Pankararé Indians in the state of Bahia captured armadillos, using their carapace to produce handicrafts such as necklaces and bracelets. They also used various species from this group for therapeutic purposes, such as the Tolypeutes tricinctus (ball armadillo), where its carapace was used to treat heart problems, rheumatism and stomach pain. The *Euphactus sexcinctus* (peba armadillo) was used to extract fat, blood and bones to treat minor injuries, while the Dasypus novemcinctus (true armadillo) had its carapace smoked and used against asthma, and its tail was also used to treat earaches and even deafness.

In the state of Alagoas, 100% of the hunters interviewed said that armadillos are one of the most hunted species because they have some of the tastiest meat (SILVA, 1993). With regard to the various uses of armadillos, it was observed that they were all related to food and are associated with cultural issues in each region. Among the most common methods used to capture these animals are firearms, hunting dogs and traps (BARBOZA, 2009).

Silva (1993) and Barboza (2009) state that the hunting of these animals is generally associated with people with low incomes and little schooling, and that these individuals depend on the use of fauna for survival, which encourages the practice.

In Brazil, various species of animals have been exploited by indigenous populations and descendants of European colonisers since the colonial period, as well as in our own time, for example when we domesticated animals such as dogs and cats (ALVES et al., 2012). This dynamic is increasing, with fauna being used for a variety of purposes, ranging from food, clothing and cultural activities to trade and smuggling of live animals, as well as for religious and medicinal rituals (ALVES et al., 2012).

According to Alves et al. (2012), in a study carried out in north-eastern Brazil, birds and mammals are the most utilised vertebrates in this region. Among the most important families is the Dasypodidae, comprising the species Euphractus sexcinctus (armadillo peba), which even feeds on the remains of other animals that are already decomposing, while the species Dasypus novemcinctus (armadillo verdadeiro), whose meat is much appreciated, is considered a clean animal because it has a diet based on roots and small insects.

As the Euphractus sexcinctus species does not have a balanced diet for consumption, it is taken to breeding centres where it is kept on a controlled diet in order to "cleanse" its digestive tract for later human consumption (ALVES et al., 2012).

3.2BILL (PL) NO. 6.268 OF 2016

In the Chamber of Deputies, there is a bill (PL) No. 6.268 of 2016, which has provisions on the National Policy compromising Brazil's wild fauna and allowing deforestation and degradation. The author of this bill cites the hunting of wild boar species, which has been allowed in the country since 2013, and his main argument is that this hunting, if allowed, could help combat exotic species. This bill may pose some risk to our ecosystem, but there is no knowledge of what the hunting of one species can do to the development and reproduction of others.

The wild boar species is being considered an "invasive pest", making it possible to release hunting for its persecution, which could bring risks to many other species that inhabit the same place. In addition, private units that adhere to the bill will have to pass on 30 per cent of their annual profits, which must be used to protect wild species in Brazil (APREMAVI, 2017).

Within this same Bill, there is also the issue of animals that harm agricultural production, in which case the armadillo will be particularly affected, as they dig holes in crops and could consequently have a large reduction in their species in Brazil.

3.3HUNTING ARMADILLOS IN WESTERN SANTA CATARINA

In a study carried out by researchers at the Federal University of Santa Catarina, members of an indigenous community in the west of Santa Catarina carried out a study on hunting, where the species soft-tailed armadillo and mulita armadillo were identified, the same ones used for food (SANTOS & SANTOS, 2015). Preuss et al. (2016), when carrying out a survey of mastofauna in the municipality of São Miguel do Oeste, in the far west of the state of Santa Catarina, observed the presence of the species Dasypus novemcinctus (chicken armadillo) and Dasypus hybridus (spider armadillo), with 11 and two records respectively in the study area of approximately 220 hectares.

When carrying out a study of roadkill in the western region of Santa Catarina, in the cities of Joaçaba, Ibicaré, Luzerna, Treze Tilhas, Água Doce and Catanduvas, Orlandin et al. (2015) found five individuals of Dasypus novemcinctus (chicken armadillo), a species that occurs on a large scale in our region.

3.4CONSERVATION OF ARMADILLO SPECIES

The exponential increase in species considered to be threatened with extinction on red lists is largely associated with practices such as hunting, road kill, habitat fragmentation and pollution (DIAS, 2013).

This makes it important to sensitise the population to the importance of conserving these species and their habitats, and environmental education is one of the tools for this. Environmental education can be understood as the process by which the student begins to learn about environmental issues and starts to take a new view of the environment, linking the axes to environmental conservation (MEDEIROS et al., 2011).

Questions about issues related to the environment are becoming more and more important in our daily lives, and are essential at all stages of educational processes in schools, with great importance in the early years of individuals' learning (MEDEIROS et al., 2011).

That's why it's necessary to propose studies to raise awareness about the importance of biodiversity conservation, making it essential to study the local fauna in greater depth (DIAS, 2013). The lack of knowledge about the biology, physiology, ecology and social perception of the species studied here makes it difficult to take specific and effective action to guarantee their conservation (ROCHA, 2009).

Knowing the importance of these listed studies, especially for animals in the process of extinction, is also a very relevant topic that should be addressed in schools, in order to increasingly increase the conservation and preservation of armadillos in their natural habitat (DIAS, 2013).

CHAPTER 4

METHODOLOGY

The study was carried out in Abelardo Luz, in the west of the state of Santa Catarina, in public schools located in the municipality. Six municipal schools were chosen, with ninth graders totalling 128 students. Four of these schools were located in rural areas and two in urban areas.

First, the project was organised and the topic and objectives of the research were defined. The project was then sent to the Brazil Platform for further evaluation of the theme and questionnaire proposed for application. Once approved, contact was made with the schools to start the research.

On the first visit to the schools, the authorisation to carry out the research was presented by the Abelardo Luz/SC Municipal Department of Education (Appendix 1) and the approval of the Brazil Platform (Appendix 2), and an authorisation signed by each school headmaster was also requested (Appendix 3). On the same day, the topic was presented to the students and they were given the Free and Informed Consent Form (Appendix 4), so that they could take it to their parents or legal guardians for them to sign, and the specific Consent Form (Appendix 5) for them to sign and agree to take part in our research and bring it signed to the appointed date.

After a few days, the signed terms were received and then a questionnaire was administered with a scale of importance on subjects relating to the habits, hunting and conservation of the Dasypodidae species (Appendix 6). The questionnaire included three questions about characterising the sample, two questions about knowledge of the genus, six questions about hunting animals and the species in question, two questions about meat used for food, 12 objective questions and one essay. The children were given approximately 30 minutes to answer the questionnaire.

After the questionnaire had been administered, a lecture lasting around 40 minutes was given, explaining the importance of the genus, its conservation and the risk of extinction to which they are exposed. A week or so after the lecture, the questionnaire was reapplied to analyse whether there had been any changes in the students' opinions on the subject.

The Excel programme was used to create the database and graphs.

CHAPTER 5

SCHEDULE

ACTIVITIES	MONTHS 2017									
	M	**A**	**M**	**J**	**J**	**A**	**S**	**O**	**N**	**D**
Presentation of the proposed topic to the students.	X	X	X							
Application of questionnaires		X	X							
Data tabulation and analysis		X	X	X						
Presentation of the partial results of TCCII					X					
Literature review	X	X	X	X	X	X	X	X	X	
TCCII adjustments						X	X			
Drafting the final version							X	X	X	
Presentation of the final results of CBT III									X	X

CHAPTER 6

RESULTS AND DISCUSSIONS

6.1AUTHORISATIONS, DELIVERY OF TERMS AND LECTURES

This study began on 24 April 2017, by contacting two schools in the urban area, E.B.M. Irineu Bornhausen with around 79 students and E.B.M. Prof° Laury Luiz Deon with 46 students. The school E.B.M. Prof° Hildo Bernardino Goulart with 13 pupils is located in the rural area of Abelardo Luz, and the first visit took place on 2nd May 2017. At the E.B.M. Prof° Ernesto Sirino school with 15 pupils, the first visit took place on 9 May 2017, also located in the rural area of the city. At the E.B.M. 25 de Maio school with 35 pupils and the E.B.M. José Maria school with 47 pupils, both located in settlements in the rural area of the city, the visits took place on 10 and 15 May 2017 respectively.

Both the urban and rural schools in the municipality received the school's authorisation on the same day as the first visit and presented the topic to the pupils, who took home the consent forms for their parents and/or guardians to sign, as well as the consent form agreeing to take part in the research. Of the 235 pupils assessed, only 128 brought the forms correctly filled in and signed by their parents and guardians.

Approximately seven days after the first visit to each school, the students were given a pre-lecture questionnaire without mentioning the subject, with the aim of assessing their previous knowledge of the genus and its conservation. On the same day after the questionnaire was administered, a lecture lasting approximately 40 minutes was given.

Around a week after the lecture, the students returned to the schools to complete the post-lecture questionnaire, which was compared with the first questionnaire to see what had changed in the students' opinion of the proposed topic, and whether the sample had been sensitised.

6.2 QUESTIONNAIRE RESULTS

Although most of the schools are located in rural areas of Abelardo Luz, we can see that the majority of the students interviewed were male (Figure 1). And according to IBGE data from the 2010 demographic census, between the two regions inhabited in the city, 27.10 per cent are men and 28.86 per cent are women in the urban area of the municipality of Abelardo Luz, while 23.35 per cent of rural residents are men and 20.7 per cent are women. Thus, it can be seen that more women live in the urban area, while in the rural area this situation is reversed, justifying the majority of male students being in the rural area of the municipality.

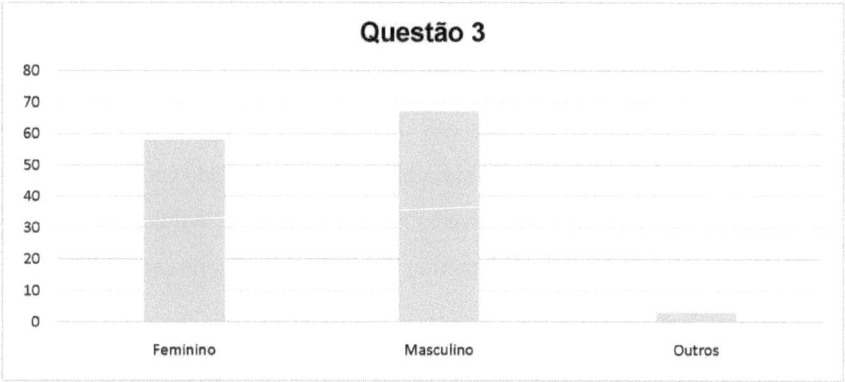

Source: GOMES, 2017.

Figura 1 Gender of the students interviewed.

Among the students interviewed, there wasn't much difference between the two age groups

(Table 1). Compared to the older students, it can be seen that the majority are within the age range for ninth graders, with only one student leaving the question blank. There was also one student, from a rural school, who was already 18 years old.

Table 1 Age of the students interviewed for the questionnaire.

Ages	Number of students	Frequency (%)
From 11 to 14 years old	75	58,6%
From 15 to 18 years old	52	41,4%

Source: GOMES, 2017.

According to the National Institute for Educational Studies and Research Anísio Teixeira (INEP, 2016), enrolment in primary schools remained highest among students aged between six and 14, indicating that the majority of students interviewed (58.6%) are within the ideal age range for primary school. Of these, the vast majority live in the rural area of the municipality (Table 2).

Table 2 Location of the homes of the students interviewed for the questionnaire.

Location	Number of students	Frequency (%)
Countryside	72	56,25
Urban	54	43,75

Source: GOMES, 2017.

According to the 2010 IBGE demographic census, 55.96% of the population lives in urban areas and 44.03% in rural areas in Abelardo Luz/SC. The greater presence of students from rural areas can be attributed to the number of schools assessed in the two regions, and of the six schools with ninth grade classes, four were located in rural areas. This data may have contributed to the fact that the vast majority of the students questioned answered that they knew the animal presented (Table 3).

Table 3 Answers when asked if they knew the animal presented (armadillo).

Reply	Number of students	Frequency (%)
YES	116	90,62
NO	12	9,38

Source: GOMES, 2017.

Despite the large number of students who know this animal, approximately 10 per cent of the students said that they don't know it and have never seen it. Of the students who said they knew the animal, the vast majority (56.25%) said they saw it once or twice a week (Figure 2). In a study by Barboza (2009), who interviewed residents of the Paraíba region to identify their perception of Xenarthras, 13.55% of those interviewed said they knew more than two types of armadillo in the region, including a rare species, the ball armadillo (Tolypeutes tricinctus).

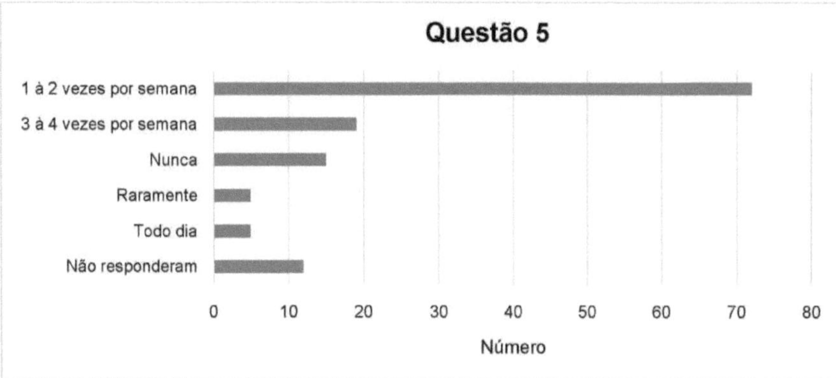

Source: GOMES, 2017.

Figura 2 Frequency with which the students interviewed had seen armadillos in their region/city.

Although there are approximately 20 species of armadillo that occur from the south of South America to the south of the United States, in the region studied by Barboza (2009), 47.7 per cent of those interviewed said they had only seen one species of armadillo, known as the true armadillo (Dasypus novemcinctus). Other interviewees said they had seen two species of armadillo in their region (38.7 per cent), the second of which was the well-known peba armadillo (E. sexcinctus).

Despite not having the alternative rarely, five students reported the need to add this alternative because they didn't always spot the animal during the week, but once a month or even once a year. On the other hand, 15 students reported that they had never seen the animal in their area and 12 left the question blank.

When asked if they knew anyone who hunted animals, 42.96 per cent of the students said they knew people and friends who did so (Figure 3). On the other hand, 46.87 per cent said they didn't know anyone who hunts. This figure may have been overestimated because these students didn't answer out of fear or shame, as they may have close relatives who hunt. It should also be noted that, among the options for family members, the father figure was the most frequently mentioned member (7.03%).

These figures are in line with those reported by Dantas & Machado (2013) who, through interviews, observed that only men took part in this practice and 100 per cent of the women interviewed reported not hunting.

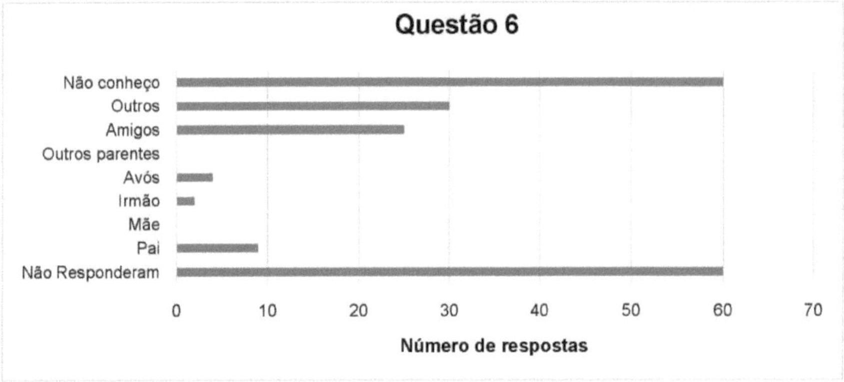

Source: GOMES, 2017.

Figura 3 Frequency of acquaintances (relatives or not) who hunt.

In addition to the previous question, it can be seen that the armadillo predominated in all the alternatives ticked by the students. In addition to the armadillo, other wild animals from our region and even others such as the wild boar and the collared peccary were ticked, but not the armadillo.

were in the alternatives (Figure 4). The large number of students who didn't answer the question may be due to the previous question where they said they didn't know anyone who hunted, so if they didn't know anyone they would leave it blank. In a study in Ceará, Boogart & Brienen (1985) reported that 36 armadillos were killed in 40 days in the region, almost one armadillo a day. In another study to assess the most commonly used food resources in settlements in the state of Pará, the armadillo and the caititu appear as the two most cited animals (CAJAÍBA et al., 2015).

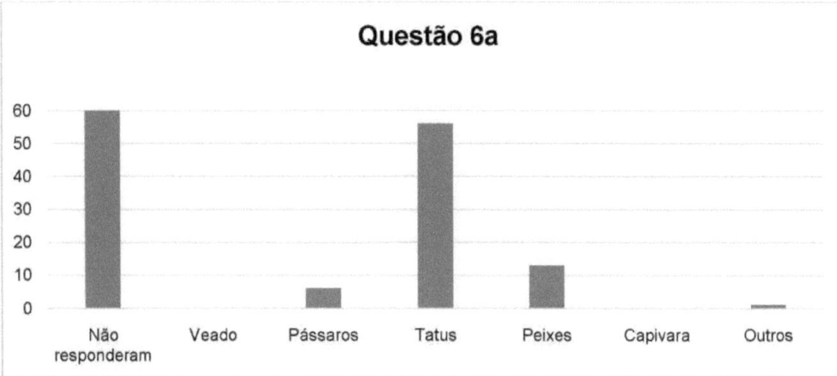

Source: GOMES, 2017.

Figura 4 Animals most hunted by acquaintances of the students interviewed.

According to IBAMA data, between 1997 and 2000 around 55 armadillos were seized. Of these, 47 were released, five went to Wild Animal Screening Centres (CETAS), two to zoos and one was killed. Compared to other mammals included in these figures, the armadillo is second only to two species of monkeys from the Primates order.

Figure 5 shows the pre- and post-lecture results, in which they were asked which animals were most hunted in our region. Again, the armadillo predominated, although other species were also ticked, as the question allowed more than one alternative to be ticked. The number of students who ticked both one and the other species didn't even reach half of those who hunted armadillos. The same result was observed by Negreiros et al. (2010), through the results of questionnaires in the state of Piauí, in which they reported that the majority of interviewees cited the armadillo as the most sought-after animal for hunting.

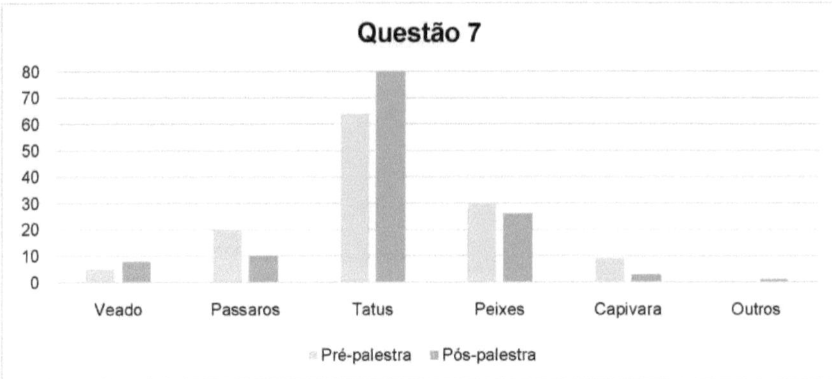

Source: GOMES, 2017.

Figura 5 The most hunted animals in the opinion of the students interviewed.

According to data from the IBGE (2001), the official list of species of Brazilian fauna threatened with extinction was made available in order to appeal to young people. Two species of armadillo that are already threatened with extinction in Brazil are Priodontes maximus (tatu-canastra, tatuaçu) and Tolypeutes tricintinus (tatu-bola, tatuapara). These two species, like others, suffer from hunting practices, trampling and deforestation, among other causes that are increasing concern for their conservation.

It can also be noted that, after being shown the correct data and information about armadillo hunting, the students reported after the lecture that armadillos are hunted more in our region/city. For the question in which students were asked to give a score from one to nine for the importance of hunting, with one for not very important and nine for very important, we noticed that there was not much difference pre- and post-lecture for each alternative (Figure 6). In a study by Negreiros et al. (2010), in which 46 open-ended questionnaires were administered to market traders, high school students and university students from two courses in the state of Piauí, they reported that the university students interviewed ranked hunting first as a leisure use, while high school students believe that their hunting is for survival, and the other groups believe that it is for other areas such as trade.

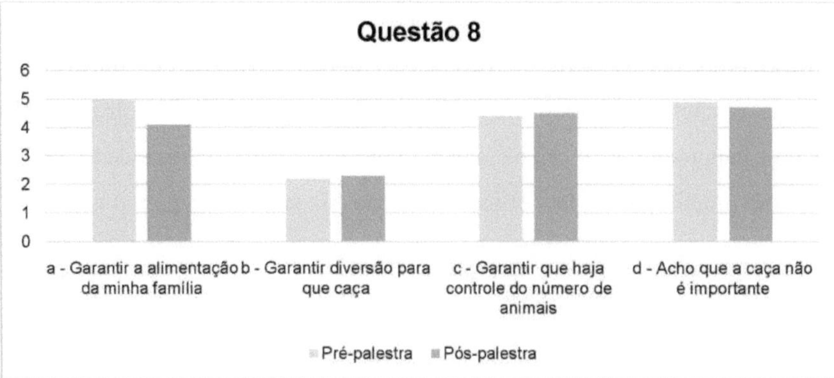

Source: GOMES, 2017.

Figura 6 Average reasons why hunting is important for students.

When the students were asked to rank the reasons why hunting should be avoided, there was a difference in the average pre- and post-lecture results (Figure 7). According to Negreiros et al. (2010), the vast majority of those interviewed said that hunting causes great destruction of biodiversity, environmental imbalance, animal extinction and affects nature. Observing that this knowledge generates damage to this practice and its main challenge is the development of a new ecological economy.

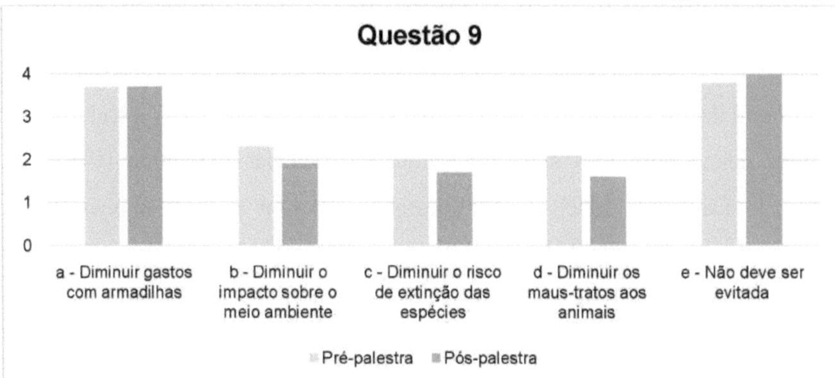

Source: GOMES, 2017.

Figura 7 Average reasons why hunting should be avoided

When asked whether the student had witnessed any kind of mistreatment or hunting of armadillos, we can see that the majority (more than a hundred students) had never witnessed such a practice, but also that the few who had witnessed it did not intervene or report it (Figure 8).

Although there is a wildlife protection law (Federal Law No. 5,197 of 3 January 1967), which protects wild fauna not specifically from mistreatment, but from hunting, trade, transport, captivity, among others, the population studied knew little about this law and how important it is. It can be seen in figure 8 that most of the students had not witnessed this practice, while few of those who had (six students) reported it.

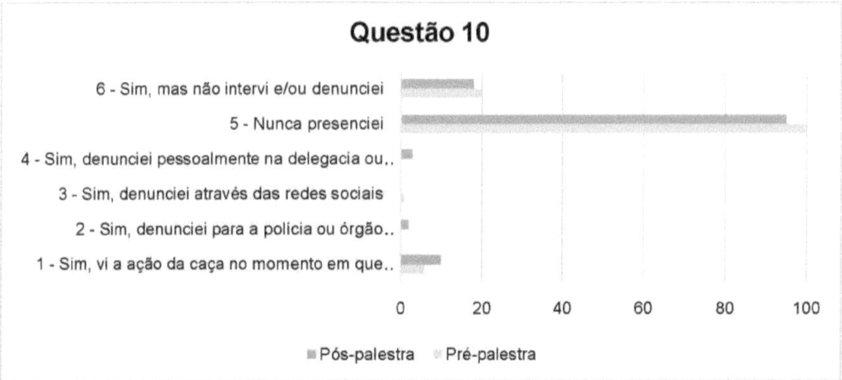

Source: GOMES, 2017.

Figura 8 Whether the students have ever witnessed any kind of mistreatment or hunting of armadillos.

Among the items that are negative effects of hunting, the one that had the most alternatives ticked both pre- and post-lecture was species extinction (Figure 9). According to Dantas & Machado (2013), the population studied emphasises the importance of preserving the ball armadillo, with 88% of those interviewed claiming that preservation is important and pointing out that its risk of extinction lies in its ecological and cultural value.

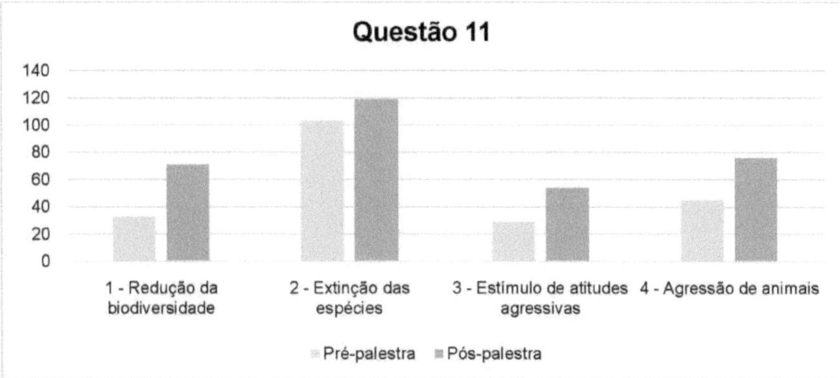

Source: GOMES, 2017.

Figura 9 Items that are bad effects of hunting on nature.

When the students were asked to tick which animals could be used without any problem for food, several wild animals were ticked. However, it is worth noting that the armadillo was ticked many times in the pre-lecture period and, after the correct presentation of the data and the risk that the species runs of extinction, its ticking decreased considerably after the lecture (Figure 10).

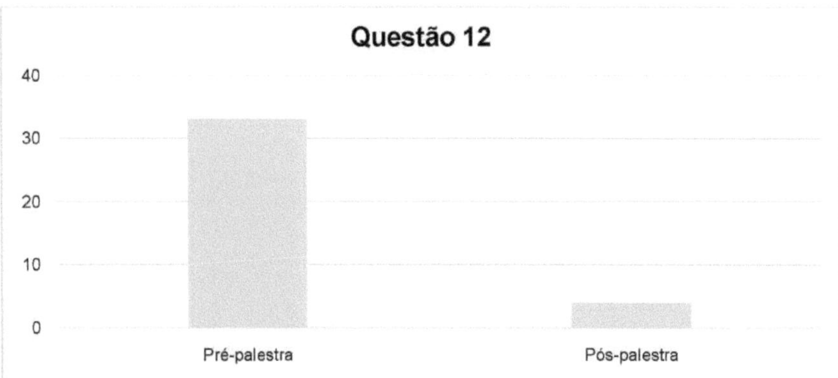

Source: GOMES, 2017.

Figura 10 Times that the armadillo has been marked that it can be used without any problem for food.

According to Negreiros et al. (2010), the habit of consuming wild game ends up occurring among all those interviewed, without their schooling interfering. However, among all the groups interviewed, the vast majority do not consume armadillo meat, followed by those who consume it rarely and, to a lesser extent, the option frequently.

CHAPTER 7

FINAL CONSIDERATIONS

The study of perception is of fundamental importance for understanding the relationship between man and the environment. When it is applied to children, they reproduce what they hear and see in their daily lives, and it is a way of expressing culture and knowledge.

When assessing the hunting of wild animals, especially the armadillo, it was possible to observe a low level of knowledge on the subject, and the species is an animal that is little discussed in schools. In addition, the armadillo suffers greatly from hunting by family and friends of the students, although comments are made for fear of jeopardising them.

On the other hand, it can be seen that, after the explanatory lecture, the students' opinions on the subject changed a lot, motivating them to apply their knowledge to reverse many thoughts about hunting wild animals. This subject is a major challenge for biology to find strategies that can reduce the rate of extinct animals influenced by humans, because the biodiversity we still have is our greatest diamond.

Therefore, the lack of knowledge about these environmental crimes indicates that alternatives need to be sought to reduce the consumption of wild animals by society.

This study provides an incentive to discuss and plan more educational activities in schools about hunting and the laws that prohibit it.

CHAPTER 8

BIBLIOGRAPHICAL REFERENCES

ALVES, R.R.N. et al.Hunting, use and conservation of vertebrates in the Brazilian semi-arid region. In: BRAZILIAN SEMINAR ON HUNTING, USE AND CONSERVATION OF VERTEBRATES, Mongabay.com Open Access Journal - **Tropical Conservation Science** v.5 n.3, p.394-416, Paraíba, 2012.

ANACLETO, T. C. S. **Distribution, diet and effects of anthropogenic changes in the Cerrado on armadillos.** 2006. 139p. Thesis (CAPES postgraduate programme) - UFG/Environmental Sciences, Goiânia.

APREMAVI - Association for the Preservation of the Environment and Life. All rights reserved. **Say no to the release of hunting in Brazil, 2017.** Available at: http://www.apremavi.org.br/diga-nao-a-liberacao-da-caca-no-brasil/ . Accessed on: 9 June 2017.

BAGER, A. **Ecology of roads.** Minas Gerais: UFLN, 2012, 374p.

BARBOZA, R. R. D.**A Etnoecologia dos tatus-peba** (Euphractus sexcinctus **(linnaeus, 1758) e tatu verdadeiro** (Dasypus novemcinctus **linnaeus, 1758) na perspectiva dos povos do semiárido paraibano.** Paraíba 2009. 100p. Dissertation (Master's Degree in Environmental Science and Technology (MCTA). State University of Paraíba.

BLOCH, A. et al. **The toothless**. In: The animals, v. 3, Bloch ed., Rio de Janeiro, 1976, 1220p.

BOOGART E.V.D. & BRIENEN R.P. Information from Ceará by Georg Marcgraf. Rio de Janeiro, 1985, 48p. **Index Publishing House**.

CAJAÍBA, R.L. et al. Wild animals used as food resources in rural settlements in the municipality of Uruará, Pará, Brazil. **DME Desenvolvimento e Meio Ambiente**, v. 34, p. 157-168. 2015.

COSTA-NETO, E. M. et al. Human/xenarthra interactions: anteaters, sloths and armadillos in Amerindian folklore. **Actual Biologia**, v.22, n.73, p. 203-213, 2000.

DANTAS, M.C. & MACHADO, D. D. **Ethnoecological study of the ball armadillo species** (Tolypeutes tricinctus) **in a community surrounding the Serra das Almas nature reserve, Ceará-Piauí.**FAEC / State University of Ceará - UECE, 2013. Available at: < http://www.revistaea.org/pf.php?idartigo=1876> Accessed on 20 September 2017.

DEPS, P.D. et al. Epidemiological aspects of leprosy transmission in relation to exposure to armadillos. **Hansen**. v.28 n.2, p. 138-144, 2003.

DIAS, V.G. The importance of studying endangered animals at school for the conservation of species based on the reports of those present at the 1st Science Fair at Escola 16 de Dezembro. **Revista Mirante,** V.3, n.1, p.83-90, 2013.

EISENBERG, J.F. & REDFORD, K.H. **Mammals of the Neotropics, v.3 - The Central Neotropics: Ecuador, Peru, Bolivia, Brazil.** The University of Chicago Press,Chicago, 624 p.1999.

HILDEBRAND M. & GOSLOW G. Mesenterial coelom. In: **Analysis of vertebrate structures**. 2 ed. Atheneu, São Paulo, 2006, p.195-199.

IBAMA - **National Campaign to Combat Trafficking in Wild Animals: final report,** Brasília, 1997-2000p. 18. Available at: <http://www.renctas.org.br/wp-content/uploads/2014/02/REL_RENCTAS_en_final.pdf> Accessed on 18 September 2017.

IBGE - Brazilian Institute of Geography and Statistics. **Demographic Census, 2010.**

Available at: http://cidades.ibge.gov.br/painel/populacao.php?lang=&codmun=420010&search=santa- catarina|abelardo-luz. Accessed on: 18 September 2017.

IBGE - Brazilian Institute of Geography and Statistics. **Endangered fauna, 2001.** Available at: https://biblioteca.ibge.gov.br/visualizacao/livros/liv775.pdf. Accessed on 20 September 2017.

INEP MEC - National Institute for Educational Studies and Research Anísio Teixeira. **Scientific Notes CENSO ESCOLAR, 2016.** Available at: http://download.inep.gov.br/educacao_basica/censo_escolar/notas_estatisticas/2017/notas_est atisticas_censo_escolar_da_educacao_basica_2016.pdf. Accessed on: 18 September 2017.

MEDEIROS, A. B. et al. The importance of environmental education at school in the early grades. **Revista Faculdade Montes Belos**, v. 4, n. 1, p 01-17, 2011.

NEGREIROS, A. B. et al. **Environmental Perception of Wild Animal Hunting: Different Views in the City of Floriano-PI.** Editora Index. p 1-8, ([2010]). Available at: <http://congressos.ifal.edu.br/index.php/connepi/CONNEPI2010/paper/viewFile/786/495> Accessed on 20 September 2017.

NOWAK, R.M. **Walker's mammals of the world Sixth edition**, The Johns Hopkins University Press, Baltimore and London, p. 147-168, London,1999.

ORLANDIN, E. et al. Medium and large-sized mammals run over in western Santa Catarina, Brazil.**Biota Amazônia - Open Journal Systems**, v. 5, n. 4, p. 125-130, 2015.

PRADA M. & MARINHO-FILHO J. Effects of fire on the abundance of Xenarthras in Mato Grosso, Brazil. **Austral Ecology**, n.29, p. 568-573, 2004.

PREUSS, J.F. et al.Rapid survey of terrestrial mammals in a remnant of Atlantic forest in southern Brazil.**Unoesc & Ciência**, v. 7, n. 1, p. 89-96, 2016.

ROCHA, E. V. Teaching environmental education with the aid of taxidermised animals. **Revista da Católica**, v. 1, n. 1, p. 201-211, 2009.

SANTOS, G. M. & SANTOS, P.R. **Práticas culturais Kaingáng na Terra Indígena Xapecó: Relatos sobre a caça e pesca na aldeia Olaria**.TRABALHO DE CONCLUSÃO DE CURSO (TCC), UNIVERSIDADE FEDERAL DE SANTA CATARINA - Santa Catarina, 2015.

SILVA, G. A. **Mammals of hunting importance in Várzea de Marituba and Fazenda Boa Vista, Alagoas: species hunted and hunting methods.** In Dissertation. Federal University of Alagoas. Alagoas, 178p, 1993.

WETZEL, R.M. The identification and distribution of recent Xenarthra (=Edentata), 5-21. In: **The evolution and ecology of armadillos, sloths and vermilinguas**, Smithisonian Institution Press, Washington and London, 451p, 1985.

CHAPTER 9

ANNEXES

9.1 ANNEX 1 - AUTHORISATION FROM THE ABELARDO LUZ/SC MUNICIPAL EDUCATION DEPARTMENT

DECLARAÇÃO

A Secretaria de Educação Municipal de Abelardo Luz-SC, por seu representante legal Pedrinho Oliveira Nunes **(Secretário Municipal de Educação)** declaro **estar ciente** da pesquisa intitulada "SENSIBILIZAÇÃO DE ALUNOS DE ESCOLAS PÚBLICAS PARA A CONSERVAÇÃO DE DASYPODIDAE NO OESTE DE SANTA CATARINA" de responsabilidade dos pesquisadores: Alan Miranda Prestes e Carla Graciele Gomes.

Em relação à pesquisa supracitada, informamos que o acesso dos pesquisadores ao local da pesquisa e a manipulação dos dados, será **autorizado somente após a emissão do PARECER DE APROVAÇÃO do Comitê de Ética em pesquisa em Seres Humanos da UNOESC/HUST, nos termos da Resolução 466/12.**

Abelardo Luz-SC

Por ser verdade, firmo a presente.

Abelardo Luz, 23 de Março de 2017.

Assinatura do Responsável pela Instituição

Pedrinho Oliveira Nunes
Responsável pela Secretaria
Municipal de Educação

9.2 ANNEX2 - APPROVAL OF THE BRAZIL PLATFORM

DETALHAR PROJETO DE PESQUISA

– DADOS DA VERSÃO DO PROJETO DE PESQUISA

Título da Pesquisa: Sensibilização de alunos de escolas públicas para a conservação de Dasypodidae no oeste de Santa Catarina
Pesquisador Responsável: ALAN MIRANDA PRESTES
Área Temática:
Versão: 3
CAAE: 65421317.3.0000.5367
Submetido em: 19/04/2017
Instituição Proponente: FUNDACAO UNIVERSIDADE DO OESTE DE SANTA CATARINA
Situação da Versão do Projeto: Aprovado
Localização atual da Versão do Projeto: Pesquisador Responsável
Patrocinador Principal: Financiamento Próprio

PLATAFORMA BRASIL COORDENADOR PLATAFORMA BRASIL

Comprovante de Recepção: PB_COMPROVANTE_RECEPCAO_875728

– DOCUMENTOS DO PROJETO DE PESQUISA

- Versão Atual Aprovada (PO) - Versão 3
  - Pendência Documental (PO) - Versão 3
    - Currículo dos Assistentes
    - Documentos do Projeto
      - Comprovante de Recepção - Submissão
      - Declaração de Instituição e Infraestrutura
      - Folha de Rosto - Submissão 5
      - Informações Básicas do Projeto - Submi
      - Outros - Submissão 5
      - Projeto Detalhado / Brochura Investigad
      - TCLE / Termos de Assentimento / Justific
    - Apreciação 5 - Universidade do Oeste de Sa
- Projeto Completo

Tipo de Documento	Situação	Arquivo	Postagem
Parecer Consubstanciado do CEP	Aceito	PB_PARECER_CONSUBSTANCIADO_CEP_2024511.pdf	20/04/2017 09:58:24

– LISTA DE APRECIAÇÕES DO PROJETO

Apreciação	Pesquisador Responsável	Versão	Submissão	Modificação	Situação	Exclusiva do Centro Coord.	Ações
PO	ALAN MIRANDA PRESTES	3	19/04/2017	20/04/2017	Aprovado	Não	

– HISTÓRICO DE TRÂMITES

Apreciação	Data/Hora	Tipo Trâmite	Versão	Perfil	Origem	Destino	Informações
PO	20/04/2017 09:58:18	Parecer liberado	3	Coordenador	Universidade do Oeste de Santa Catarina - UNOESC	PESQUISADOR	
PO	20/04/2017 09:57:37	Parecer do colegiado emitido	3	Coordenador	Universidade do Oeste de Santa Catarina - UNOESC	Universidade do Oeste de Santa Catarina - UNOESC	
PO	20/04/2017 09:56:12	Parecer do relator emitido	3	Coordenador	Universidade do Oeste de Santa Catarina - UNOESC	Universidade do Oeste de Santa Catarina - UNOESC	
PO	20/04/2017 09:47:32	Aceitação de Elaboração de Relatoria	3	Coordenador	Universidade do Oeste de Santa Catarina - UNOESC	Universidade do Oeste de Santa Catarina - UNOESC	
PO	20/04/2017 09:46:44	Confirmação de Indicação de Relatoria	3	Coordenador	Universidade do Oeste de Santa Catarina - UNOESC	Universidade do Oeste de Santa Catarina - UNOESC	
PO	20/04/2017 09:45:23	Indicação de Relatoria	3	Coordenador	Universidade do Oeste de Santa Catarina - UNOESC	Universidade do Oeste de Santa Catarina - UNOESC	
PO	20/04/2017 09:44:45	Aceitação do PP	3	Coordenador	Universidade do Oeste de Santa Catarina - UNOESC	Universidade do Oeste de Santa Catarina - UNOESC	
PO	19/04/2017 23:14:02	Submetido para avaliação do CEP	3	Pesquisador Principal	PESQUISADOR	Universidade do Oeste de Santa Catarina - UNOESC	
PO	19/04/2017 13:58:39	Rejeição do PP	3	Coordenador	Universidade do Oeste de Santa Catarina - UNOESC	PESQUISADOR	Prezado pesquisador,Recebemos seu e-mail esclareceVer mais >>
PO	18/04/2017 21:55:04	Submetido para avaliação do CEP	3	Pesquisador Principal	PESQUISADOR	Universidade do Oeste de Santa Catarina - UNOESC	

Ocorrência 1 a 10 de 29 registro(s) » »»

9.3 ANNEX 3 - AUTHORISATION FROM MUNICIPAL SCHOOL HEADMASTERS

DECLARAÇÃO

A E.A.M. IRINEU BORNHAUSEN de Abelardo Luz/SC, por seu representante legal Cassiane dos Santos Nunes (**Diretor(a)**declaro **estar ciente** da pesquisa intitulada"SENSIBILIZAÇÃO DE ALUNOS DE ESCOLAS PÚBLICAS PARA A CONSERVAÇÃO DE DASYPODIDAE (TATU) NO OESTE DE SANTA CATARINA"de responsabilidade do(s)pesquisador(es): Alan Miranda Prestes e Carla Graciele Gomes.

Em relação à pesquisa supracitada, informamos que o acesso dos pesquisadores ao local da pesquisa e a manipulação dos dados, será **autorizado somente após a emissão do PARECER DE APROVAÇÃO do Comitê de Ética em pesquisa em Seres Humanos da UNOESC/HUST, nos termos da Resolução 466/12.**

Abelardo Luz-SC

Por ser verdade, firmo a presente.

Abelardo Luz, 26... de ...abril.........de 2017

E.A.M IRINEU BORNHAUSEN
Cassiane Dos Santos Nunes
Diretora da Escola
Mat nº - 4216

Cassiane dos Santos Nunes

Assinatura do Responsável pela Instituição

DECLARAÇÃO

A E.B.M. Profº LAURY LUIZ DEON de Abelardo Luz/SC, por seu representante legal ______________________ **(Diretor(a)**declaro **estar ciente** da pesquisa intitulada"SENSIBILIZAÇÃO DE ALUNOS DE ESCOLAS PÚBLICAS PARA A CONSERVAÇÃO DE DASYPODIDAE (TATU) NO OESTE DE SANTA CATARINA"de responsabilidade do(s)pesquisador(es): Alan Miranda Prestes e Carla Graciele Gomes.

Em relação à pesquisa supracitada, informamos que o acesso dos pesquisadores ao local da pesquisa e a manipulação dos dados, será **autorizado somente após a emissão do PARECER DE APROVAÇÃO do Comitê de Ética em pesquisa em Seres Humanos da UNOESC/HUST, nos termos da Resolução 466/12.**

Abelardo Luz-SC

Por ser verdade, firmo a presente.

Abelardo Luz, 26 de Abril de 2017

____

Assinatura do Responsável pela Instituição

E.B.M. PROF. LAURY LUIZ DEON
LEI Nº 1.716/2006
Rua Renato Adolfo Begnini, 683
Centro - Abelardo Luz - SC
CEP 89830-000 - Fone (49) 3445-4910

DECLARAÇÃO

A E.B.M. Profª Hildo B. Goulart (Vila Alegre do Marco) de Abelardo Luz-SC, por seu representante legal Giovana Fiorese (Diretor(a)) declaro **estar ciente** da pesquisa intitulada "SENSIBILIZAÇÃO DE ALUNOS DE ESCOLAS PÚBLICAS PARA A CONSERVAÇÃO DE DASYPODIDAE (TATU) NO OESTE DE SANTA CATARINA" de responsabilidade do(s) pesquisador(es): Alan Miranda Prestes e Carla Graciele Gomes.

Em relação à pesquisa supracitada, informamos que o acesso dos pesquisadores ao local da pesquisa e a manipulação dos dados, será **autorizado somente após a emissão do PARECER DE APROVAÇÃO do Comitê de Ética em pesquisa em Seres Humanos da UNOESC/HUST, nos termos da Resolução 466/12.**

Abelardo Luz-SC

Por ser verdade, firmo a presente.

Abelardo Luz, 05 de Maio de 2017

GIOVANA FIORESE
DIRETORA DA ESCOLA
DEC 110/2017 Mat.:4217
E.B.M. PROF HILDO B. GOULART

Assinatura do Responsável pela Instituição

DECLARAÇÃO

A E.B.M. Prof ° ERNESTO SIRINO (Araça) de Abelardo Luz/SC, por seu representante legal Tatiani P. Tasca Stefani (**Diretor(a)**)declaro **estar ciente** da pesquisa intitulada"SENSIBILIZAÇÃO DE ALUNOS DE ESCOLAS PÚBLICAS PARA A CONSERVAÇÃO DE DASYPODIDAE (TATU) NO OESTE DE SANTA CATARINA"de responsabilidade do(s)pesquisador(es): Alan Miranda Prestes e Carla Graciele Gomes.

Em relação à pesquisa supracitada, informamos que o acesso dos pesquisadores ao local da pesquisa e a manipulação dos dados, será **autorizado somente após a emissão do PARECER DE APROVAÇÃO do Comitê de Ética em pesquisa em Seres Humanos da UNOESC/HUST, nos termos da Resolução 466/12.**

Abelardo Luz-SC

Por ser verdade, firmo a presente.

Abelardo Luz, 09 de maio de 2017

E.B.M. ERNESTO SIRINO
Tatiane P. Tasca Stefani
DIRETORA
MAT. 199602

Assinatura do Responsável pela Instituição

DECLARAÇÃO

A E.B.M. 25 de Maio de Abelardo Luz-SC, por seu representante legal Elizangela Ap Scheis **(Diretor(a))** declaro **estar ciente** da pesquisa intitulada "SENSIBILIZAÇÃO DE ALUNOS DE ESCOLAS PÚBLICAS PARA A CONSERVAÇÃO DE DASYPODIDAE (TATU) NO OESTE DE SANTA CATARINA" de responsabilidade do(s) pesquisador(es): Alan Miranda Prestes e Carla Graciele Gomes.

Em relação à pesquisa supracitada, informamos que o acesso dos pesquisadores ao local da pesquisa e a manipulação dos dados, será **autorizado somente após a emissão do PARECER DE APROVAÇÃO do Comitê de Ética em pesquisa em Seres Humanos da UNOESC/HUST, nos termos da Resolução 466/12.**

Abelardo Luz-SC

Por ser verdade, firmo a presente.

Abelardo Luz, 10 de Maio de 2017

Assinatura do Responsável pela Instituição

E.B.M 25 de Maio
ELIZANGELA A. SCHEIS
Diretora
Mat. nº 4228 - Dec. nº 119/2017

DECLARAÇÃO

A E.B.M. JOSÉ MARIA de Abelardo Luz-SC, por seu representante legal Edinéia de Fatima Fortuna (**Diretor(a)** declaro **estar ciente** da pesquisa intitulada "SENSIBILIZAÇÃO DE ALUNOS DE ESCOLAS PÚBLICAS PARA A CONSERVAÇÃO DE DASYPODIDAE (TATU) NO OESTE DE SANTA CATARINA" de responsabilidade do(s) pesquisador(es): Alan Miranda Prestes e Carla Graciele Gomes.

Em relação à pesquisa supracitada, informamos que o acesso dos pesquisadores ao local da pesquisa e a manipulação dos dados, será **autorizado somente após a emissão do PARECER DE APROVAÇÃO do Comitê de Ética em pesquisa em Seres Humanos da UNOESC/HUST, nos termos da Resolução 466/12.**

Abelardo Luz-SC

Por ser verdade, firmo a presente.

Abelardo Luz, 15 de Maio de 2017

Edinéia de Fatima Fortuna

Assinatura do Responsável pela Instituição

E.B.M JOSÉ MARIA
EDINÉIA F. FORTUNA
Diretora
Mat. nº 4096

9.4 ANNEX 4 - TCLE FOR PARENTS OR GUARDIANS

TERM OF FREE AND INFORMED CONSENT (TCLE)
FOR PARENTS OR GUARDIANS
Research title: **SENSITISATION OF PUBLIC SCHOOL STUDENTS TO THE CONSERVATION OF DASYPODIDAE IN THE WEST OF SANTA CATARINA**
Researcher responsible: Alan Miranda Prestes
Contact telephone number(s): (55) 99637-7910 or (49) 99820-0339
E-mail: alanprestes_sm@hotmail.com Total research period: 17/04/2017 to 17/05/2017

1. Your child.. is being invited to
2. the purpose of the research is to gather opinions on the conservation of wild species, especially the armadillo, from primary school pupils. Their participation is justified by the need to investigate the students' knowledge of the subject and the way in which the topic is approached in school classrooms; 3. Participation in this research will involve the following data collection procedure(s): a questionnaire will be administered (after approval by those responsible) with questions relating to the students' knowledge of the species of armadillos that exist in the region. After applying the questionnaire, a presentation and awareness-raising talk will be given to the students. Finally, there will be a new questionnaire to find out if the students have understood the subject;
4. There are no foreseeable risks or discomforts in your child's participation.
5. Your child's participation will not lead to any prejudice, discrimination or social inequality;
6. The possible benefits of the research are to raise students' awareness of the issue;
7. The results of this study may be published, but your child's name or identification will not be revealed;
8. There will be no remuneration or allowance (reimbursement) for participation
9. Any questions you may have regarding the research or your child's participation, before or after consent, will be answered by Alan Miranda Prestes, telephone: (55) 99637-7910, Rua Antônio Vitorio Giordani, 851, Xanxerê-SC and Carla Graciele Gomes, telephone: (49) 99820-0339^ rua Pedro Ivo Campo, 1706, Bairro Santa Luzia, Abelardo Luz-SC; or the Unoesc/Hust Human Research Ethics Committee, telephone: (49) 3551-2012.
10. Thus, this term is in accordance with Resolution 466/2012 of the National Health Council, to protect the rights of human beings in research. If you have any doubts about your child's rights as a research subject, or if you feel that your child has been put at unintended risk, you can contact the Research Ethics Committee for clarification;

I have read the above information and received explanations about the nature, risks and benefits of the project. I agree to my child's participation and understand that I can withdraw my consent and stop the project at any time, without penalty or loss of benefit.
Signature of the parent or guardian or **fingerprint.**
Dactyloscopic printing
Signature:
Name legible:
Address:
RG.
Phone:
Date _____ ______ /I _____

Signature of the researcher responsible
_________ Date//

9.5 ANNEX 5 - TCLE CONSENT FORM FOR STUDENTS

Term of Assent (TA)
Children aged 07 to 11 and/or teenagers aged 12 to 17
You are being invited as a volunteer to take part in the research project **"AWARENESS OF STUDENTS IN PUBLIC SCHOOLS TO THE CONSERVATION OF DASYPODIDAE IN THE WEST OF SANTA CATARINA".** In this study we intend to survey students' knowledge of the subject and raise community awareness of wildlife conservation.
The reason we are studying this subject is to gather opinions on the conservation of wild species, especially the armadillo, from primary school pupils. The research is justified by the need to investigate students' knowledge of the

subject and the way in which it is approached in school classrooms.
For this study we will adopt the following procedure(s): a questionnaire will be administered (after approval from those responsible) with questions relating to the students' knowledge of the species of armadillos that exist in the region. After applying the questionnaire, a presentation and awareness-raising talk will be given to the students. Finally, there will be a new questionnaire to find out if the students have understood the subject.
To take part in this study, your guardian must authorise you to do so and sign a consent form. There will be no cost to you, nor will you receive any financial advantage. You will be informed in any way you wish and you are free to participate or refuse. Your carer can withdraw consent or stop your participation at any time. Your participation is voluntary and refusal to take part will not lead to any penalty or change in the way you are treated by the researcher, who will treat your identity with professional standards of confidentiality. You will not be identified in any publication. This study poses no risk to your health. However, you have the right to compensation or indemnity in the event of any damage caused by the research.
The results will be available to you when finalised. Your name or any material indicating your participation will not be released without the permission of the person responsible for you. The data and instruments used in the research will be kept on file with the researcher responsible for a period of 5 years, after which time they will be destroyed. This consent form is printed in two copies, one of which will be kept by the researcher responsible and the other will be given to you.
I, ____, bearer of the identity document of
Identity (if I already have one), I have been informed of the objectives of the programme.
this study in a clear manner. I know that at any time I can request further information from the researcher in charge, **Alan Miranda Prestes, telephone: (55) 99637-7910, Rua Antônio Vitorio Giordani, 851, Xanxerê-SC,** or from the academic, **Carla Graciele Gomes, telephone: (49) 99820-0339, Rua Pedro Ivo Campo, 1706, Bairro Santa Luzia, Abelardo Luz-SC**, or from the Human Research Ethics Committee of Unoesc and Hust, telephone (49) 3551-2012. I am aware that my guardian will be able to change my decision to take part in the research if they so wish. With my guardian's consent already signed, I declare that I agree to take part in this study. I have received a copy of this consent form and have been given the opportunity to read it and clarify my doubts.
___ of 2017.

Signature of the minor or fingerprint.
Dactyloscopic printing
Name legible:
Address: RG.
Telephone:
Date______ / ____ / _____

Signature of the researcher responsible
Name legible:
Address:
Telephone:
Date______ _____ //_____

9.6 ANNEX 6 - QUESTIONNAIRE

(Mark your chosen alternative with an X)

1- Age

( ) 11 to 14 ( ) 15 to 18

2- Region where you live

( ) Rural ( ) Urban

3- Gender

( ) Female ( ) Male ( ) Other

4- Do you know what animal this is?

( ) Yes, his name is __________ ( ) No

5- Have you ever seen or heard of the animal mentioned above in your region/city?

( ) Yes ( ) No

When?

( )Every day ( ) 1 to 2 times a week

( ) 3 to 4 times a week ( ) Never

6- Do you know anyone who hunts animals? Who?

( ) Father ( ) Grandparents

( ) Mother ( ) Other relatives

( ) Siblings ( ) Friends

( ) Other, who? ______________________

( ) I don't know

6a.If so, what animals does this person hunt?

( ) Deer ( ) Birds ( ) Armadillos

( ) Fish ( ) Capybaras

( ) Others? Which? ________________________________

7- Which animals do you think are most hunted in our region/city?

( ) Deer ( ) Birds ( ) Armadillos

( ) Fish ( ) Capybaras

( ) Others? Which? ________________________________

8- Give a score from 1 to 9 for the reasons why you think hunting is important, 1 for not very important and 9 for very important:

a- Guaranteeing my family's food ( )

b- Ensuring fun for those who hunt ( )

c- Ensure that the number of animals is controlled ( )

d- I don't think hunting is important ( )

9- Rank the reasons why hunting should be avoided from 1 to 5, with 1 being the most important and 5 the least important.

( ) Reduce spending on traps

( ) Reduce the impact on the environment

( ) Reduce the risk of species extinction]

( ) Reduce the mistreatment of animals

( ) Should not be avoided

10- Have you ever witnessed armadillos being mistreated or hunted?

( ) Yes, I saw the hunting action as it was happening.

( ) Yes, I reported it to the police or the organisation responsible.

( ) Yes, I reported it via social media

( ) Yes, I personally reported it to the police station or environmental agency

( ) Never witnessed

( ) Yes, but I didn't intervene and/or report it

11- Which of the following do you think are bad effects of hunting on nature? A - ( ) Reduction in biodiversity

B - ( ) Extinction of species

C - ( ) Stimulation of aggressive attitudes

D - ( ) Animal aggression

12- In your opinion, mark with an "X" the animals listed below that could be used without any problems for our food:

Note: You can tick more than one alternative.

( ) Pig () Cow ( ) Chicken

( ) armadillo () duck ( ) rabbit

( ) Capybara () Toad ( ) Snake

( ) Sheep   () Horse

13- Write next to each type of meat how many times a week you usually eat each one:

( ) White meat (fish, frogs, turtles, alligators, birds...)

( ) Red meat (beef, pork, cat, dog, armadillo, jaguar)

( ) I don't eat meat

10 APPENDICES

10.1 APPENDIX 1 - SCIENTIFIC ARTICLE

10.2

MIX
Papier aus verantwortungsvollen Quellen
Paper from responsible sources
FSC® C105338

Printed by Books on Demand GmbH, Norderstedt / Germany